CORE
CHEMISTRY

Acids
and Alkalis

DENISE WALKER

A⁺
Smart Apple Media
an imprint of Black Rabbit Books

This book has been published in cooperation with Evans Publishing Group.

Series editor: Harriet Brown, Editor: Harriet Brown, Design: Simon Morse, Illustrations: Ian Thompson, Simon Morse

Published in the United States by Smart Apple Media
2140 Howard Drive West
North Mankato, Minnesota 56003

Library of Congress Cataloging-in-Publication Data

Walker, Denise.
Acids and alkalis / by Denise Walker.
p. cm. – (Core chemistry)
Includes index.
ISBN 978-1-58340-821-6
1. Acids. 2. Alkalies I. Title.

TP213.W35 2007
546'.24—dc22   2007001250

9 8 7 6 5 4 3 2 1

Contents

Introduction

Acids and alkalis are found in a surprising number of places. Some acids and alkalis are edible and are found in foods. Others are very strong and can be harmful,

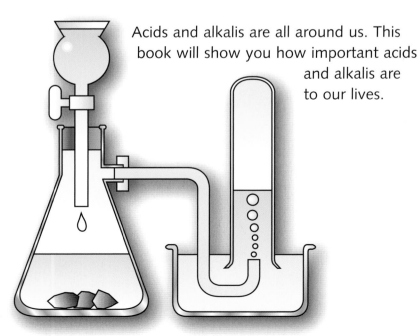

such as the acid in car batteries and the alkali in oven cleaners.

This book invites you to discover more about acids and alkalis. Find out what they are, how they affect other substances, and why we need them. Discover how acid rain forms and ways that it affects the environment. Learn about acid and alkali indicators and make your own indicator out of cabbage leaves. You will also find out how acids and alkalis are used in fire extinguishers; discover how neutralization reactions work; and even use this knowledge to stop bee and wasp stings from hurting.

This book contains feature boxes that will help you understand more about acids and alkalis. Test yourself on what you are learning; investigate some of the concepts discussed; find out key facts; and discover some of the scientific findings of the past and how these might be utilized in the future.

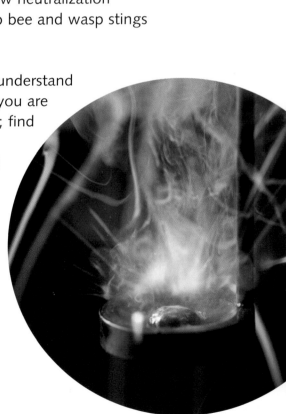

Acids and alkalis are all around us. This book will show you how important acids and alkalis are to our lives.

Did you know?

▶ Look for these boxes. They contain interesting facts about acids and alkalis in the world around us.

Test yourself

▶ Use these boxes to see how much you've learned. Try to answer the questions without looking at the book, but take a look if you are really stuck.

Investigate

▶ These boxes contain experiments that you can carry out at home. The equipment you need is usually inexpensive and easy to find around the house.

Time travel

▶ These boxes describe scientific discoveries from the past and fascinating developments that pave the way for the advance of science in the future.

Answers

On pages 46 and 47, you will find the answers to the questions from the "Test yourself" and "Investigate" boxes.

Glossary

Words highlighted in **bold** are described in detail in the glossary on pages 46 and 47.

What are acids and alkalis?

The term **acid** is often associated with a dangerous substance that can cause harm. Most concentrated acids are dangerous and **corrosive** and must be handled with care. The same is true of concentrated **alkalis**. But acids and alkalis are not always dangerous. For example, lemon juice and vinegar are acidic, and sodium bicarbonate—or baking soda—is alkaline.

WHAT IS AN ACID?

An acid is defined as a substance that **dissolves** in water and forms hydrogen **ions** (H^+). An acid can also be defined as a substance that dissolves in water and forms protons. A hydrogen atom contains one proton in its nucleus and has one electron circling the nucleus. When the hydrogen atom loses its electron to form a hydrogen ion, this leaves behind one proton. So a hydrogen ion is sometimes called a proton.

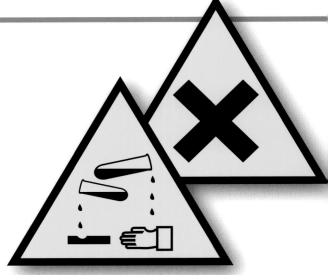

▲ The warning triangle on the right means that the substance involved is an irritant. The one on the left means that the substance is corrosive.

HYDROGEN ATOM   **HYDROGEN ION**

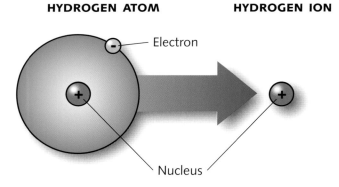

Electron

Nucleus

PROPERTIES OF ACIDS

▶ Acids can be corrosives or **irritants**. When acids are concentrated, they burn bare skin. Scientists must be careful and wear gloves when using these acids. Less concentrated acids are not as harmful and are described as irritants. These acids cause an itching sensation if spilled on bare skin.

▶ All acids have a sour taste. Some acids are present in foods that are used for flavoring. Vinegar contains acetic acid. Lemons and other citrus fruits contain citric acid. Vinegar and citrus fruits have a sharp taste which we describe as "sour." All acids have a sour taste, but you should never test this for yourself.

▶ All acids must be dissolved in water in order for hydrogen ions to form.

▶ All acids will change the color of another chemical substance called an **indicator**. The presence of an acid can be detected by adding an indicator.

USEFUL ACIDS

▶ The proteins in our bodies are made from smaller units called amino acids.

▶ Vitamin C is another name for ascorbic acid. Our bodies need this vitamin to stay healthy. Vitamin C also prevents a disease called scurvy.

▶ Salicylic acid is used to make aspirin.

▶ The acid found in car batteries is called sulfuric acid. This is a corrosive acid, which is why car batteries must be handled with care.

▶ The food that we eat is digested in our stomachs in the presence of hydrochloric acid. This is also a corrosive acid, which is why our stomachs are heavily lined with tissue that is not easily corroded.

▶ One of the most corrosive acids is called hydrofluoric acid. This acid is so corrosive that it will chemically attack a glass bottle it is stored in. Because of this, hydrofluoric acid must be stored in a plastic container. Hydrofluoric acid is used in glass etching.

▲ Citrus fruits contain ascorbic acid. This is not concentrated, so it will not damage your mouth when you eat it.

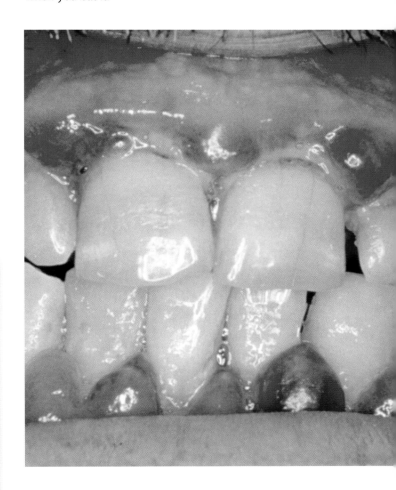

▲ These gums are swollen and inflamed because the patient has a vitamin C (ascorbic acid) deficiency and has developed scurvy.

INVESTIGATE

▶ **You must ask an adult for permission before carrying out the following experiments**.

Experiment 1
Take a chicken bone and immerse it in vinegar overnight. What happens to the bone?

Experiment 2
You will need a fresh egg, a bottle with a neck that is slightly smaller than the egg, and vinegar. Soak the egg in the vinegar for approximately two days. Carefully squeeze the egg into the bottle. Does it go all the way in? How does this work?

Acid rain

Vehicles burn fuel and emit a gas called sulfur dioxide (SO_2) into the atmosphere. Sulfur dioxide is a colorless gas that has a pungent smell. It is the main cause of **acid rain**, although nitrogen oxides from burning fuel also contribute to acid rain. Sulfur dioxide gas rises into the atmosphere and dissolves in rain clouds to form a weakly acidic solution. A series of chemical reactions in the clouds produces sulfuric acid (H_2SO_4), or acid rain.

What happens when acid rain falls to the earth?

When the acid rain falls to the earth it can:

▶ **damage buildings made from sedimentary rock.**
When acid reacts with calcium carbonate, or limestone, the rock wears away. The chemical equation for this reaction is as follows:

Calcium carbonate + Sulfuric acid $\longrightarrow$ Calcium sulfate + Water + Carbon dioxide

$$CaCO_{3(s)} + H_2SO_{4(aq)} \longrightarrow CaSO_{4(aq)} + H_2O_{(l)} + CO_{2(g)}$$

▶ **make ponds and lakes more acidic.**
Many living organisms are sensitive to these changes and cannot easily survive in acidic conditions.

▶ **strip away tree bark, which can kill the tree.**
Whole forests can be destroyed if acid rain falls in large enough quantities.

▶ **damage metallic constructions.**
Acid rain chemically attacks metal and produces a salt and hydrogen gas. The chemical equation for this reaction is as follows:

Iron + Sulfuric acid $\longrightarrow$ Iron (II) sulfate + Hydrogen

$$Fe_{(s)} + H_2SO_{4(aq)} \longrightarrow FeSO_{4(aq)} + H_{2(g)}$$

▶ This forest in Poland has been destroyed by acid rain. It is likely that the sulfur dioxide and nitrogen oxides were released by vehicles and factories thousands of miles from the forest. The gases would have been carried in the atmosphere by the wind until they dissolved in rain clouds and fell as acid rain.

WHAT ARE BASES AND ALKALIS?

Bases are compounds that can neutralize acids. When they react with acids in appropriate proportions, a neutral product is produced. The product of a **neutralization** reaction is called a **salt**. Bases are usually metal oxides, metal hydroxides, metal carbonates, or metal hydrogencarbonates. Ammonia is also a base and has the formula NH_3.

Many bases are insoluble in water. When a base is soluble in water, it is called an alkali. When a base dissolves in water to form an alkali, it releases hydroxide ions (OH^-). In summary:

▶ An alkali is a base that is soluble in water.
▶ All alkalis are bases.
▶ Only soluble bases are alkalis.

PROPERTIES AND USES OF BASES AND ALKALIS

Some bases and alkalis can be corrosive at high concentrations. They must be treated as carefully as acids. Bases have a bitter, rather than a sour taste, but you should never taste them. Bases and alkalis feel soapy to the touch, although it is dangerous to touch them because they can burn the skin.

Bases and alkalis neutralize acids. They are often used in cleaning products. Alkalis and bases are important for certain chemical reactions.

For example, some fire extinguishers contain acids and hydrogencarbonates. When the two chemicals react, carbon dioxide gas is produced. As the pressure of the carbon dioxide gas builds, it is expelled from the extinguisher at high speed to smother a fire.

▼ This electrical fire is extinguished with carbon dioxide that is produced by reacting an acid and a base.

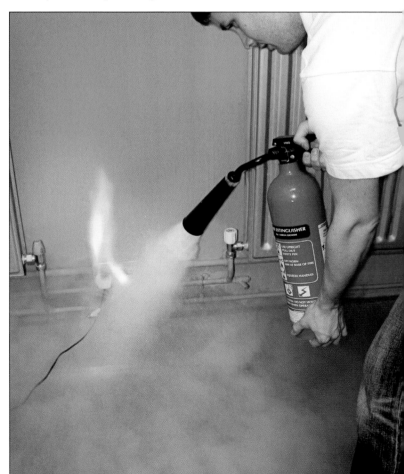

Measuring acidity and alkalinity

How do we know if a substance is an acid or an alkali? For safety reasons, we cannot taste or touch the substance, so we must use chemical testing. One of the simplest methods of determining whether a substance is acidic or alkaline is to use an indicator. An indicator is a substance that will change color in the presence of an acid or an alkali.

LITMUS

Litmus is one of the most commonly used indicators. It is available in two colors—blue litmus and red litmus. It can be used in a paper form or in solution. When you add an acid to litmus, the indicator turns (or remains) red. When you add an alkali, the indicator turns (or remains) blue. Litmus is extracted from plants called lichens that grow on rocks and trees. Scientists have used litmus as an indicator for more than 400 years.

Adding an acid or an alkali to litmus produces a chemical change that can only be reversed by adding the opposite substance. If you have added an acid to litmus, you can restore the original color by adding an alkali.

Use the following to help you remember the color changes of red and blue litmus:
AciDs turn litmus reD. ALkalis turn litmus bLue.

Red litmus	Blue litmus
Red in acid	Red in acid
Blue in alkali	Blue in alkali

► These orange and white lichens are growing on a rock. Lichens are used to make litmus indicators.

OTHER INDICATORS

Natural indicators are extracted from plants, but chemists have also developed chemical indicators. The following table shows some common indicators and the color changes they show in acids and alkalis. Notice that their chemical names give no indication of their origins, but it is worth remembering that many come from plants.

Indicator	Color in acid	Color in alkali
Methyl orange	Red	Yellow
Methyl red	Red	Yellow
Phenolphthalein	Colorless	Pink

▲ Phenolphthalein is colorless in acid and pink in alkali (left). Methyl orange is red in acid and yellow in alkali (right).

INVESTIGATE

▶ **Ask an adult's permission before trying the following experiment.**

The goal of this experiment is to make an indicator from common kitchen and garden ingredients and test it using some kitchen acids and alkalis.
To make the indicator, follow these steps:
(1) Chop red cabbage and boil it in a saucepan of water until the color of the water no longer changes—it will be purple.
(2) Place a strainer over another pan and pour the cabbage into the strainer, saving the purple-colored water.
(3) Discard the cabbage.
(4) Allow the purple water to cool.

You must wear protective clothing, rubber gloves, and goggles for the next part of the experiment.

To use the indicator:
(1) Ask an adult to help you pour household ammonia or oven cleaner into a glass.
(2) Pour lemon juice into a second glass.
(3) Add a small amount of the prepared indicator (purple water) to each of the liquids.
(4) Observe the color changes.
Which color indicates the substance is an alkali, and which color indicates an acid?

STRONG AND WEAK ACIDS AND ALKALIS

Chemists describe an acid as either strong or weak. All acids contain hydrogen. When acids break down, they release hydrogen ions. Substances that readily do this are strong acids, and those that do not easily release hydrogen ions are called weak acids.

The strength of an acid is not related to its concentration, but to its ability to decompose and release hydrogen ions. For example, the acid in vinegar is acetic acid. This is a weak acid. No matter how much acid is present, or how concentrated it is, the acid will not easily release hydrogen ions. Also, vinegar will not cause damage if it comes in contact with human skin.

▼ This beaker contains a very strong acid called hydrochloric acid. It is a volatile liquid—it **evaporates** easily—which is why you can see vapor rising from the surface.

The same principle is applied to an alkali. Alkalis contain hydroxide ions, and strong alkalis break down to release many of these ions. Weak alkalis will not break down as readily.

THE pH SCALE

A definite measure of acidity or alkalinity is found using the pH scale. The pH unit represents the "power of hydrogen." The scale runs from pH 1 to pH 14. A neutral substance has a pH value of 7. Numbers lower than pH 7 indicate an acid, with pH 1 indicating the strongest acids, such as hydrochloric acid. Very weak acids have pH values of around 5 or 6. Acid rain is a weak acid.

A value higher than pH 7 indicates that the substance is alkaline. Strong alkalis, such as sodium hydroxide, have pH values between 12 and 14. Values between pH 8 and 9 indicate very weak alkalis. Seawater is a weak alkali.

pH SCALE

pH values	Examples
pH 1	Battery acid
pH 2	Sulfuric acid
pH 3	Vinegar
pH 4	Orange juice
	Acid rain (4.2–4.4)
	Acidic lake (4.5)
pH 5	Bananas (5.0–5.5)
	Clean rain (5.6)
pH 6	Healthy lake (6.5)
	Milk (6.5–6.8)
pH 7	Pure water
pH 8	Seawater
pH 9	Baking soda
pH 10	Milk of magnesia
pH 11	Ammonia
pH 12	Sodium hydroxide
pH 13	Bleach
pH 14	Liquid drain cleaner

MEASURING pH

Simple indicators change color at specific pH values. However, not all indicators change color at the same pH. A very useful indicator that changes color over the entire range of pH values is called a **universal indicator**. This substance is a mixture of several different indicators, all of which have specific color changes at particular pH values. In acidic substances, the universal indicator displays a range of colors from red to yellow. In alkaline substances, the universal indicator can be any color from green to dark purple.

A universal indicator is available as a solution or in paper form. Solution works best if you have a colorless substance to test, such as seawater. You can add drops of the solution to the substance and observe the color change. To use paper, place a strip of the paper on a substance, such as a bar of soap. The paper changes color. It is best to use paper if you have a dark-colored substance to test. If you used indicator solution in a dark liquid, you would not see the color change.

UNIVERSAL INDICATOR

| 1 | 2 | 3 | 4 | 5 | 6 | 7 | 8 | 9 | 10 | 11 | 12 | 13 | 14 |

Strong acids   Weak acids   Weak alkalis   Strong alkalis

Neutral

USING TECHNOLOGY TO MEASURE pH

A pH probe can also be used to measure the pH. The probe consists of a glass bulb that contains a thin strip of platinum wire. Platinum is used because it is chemically unaffected by any hydrogen ions in the test solution. Before the probe is dipped into the test solution, it is calibrated using a **buffer solution** of known pH value. By doing this, you make sure that the probe is accurately measuring the pH value.

When the probe is dipped into an acidic or alkaline test solution, hydrogen ions move into the glass bulb. The probe converts this into a pH value, that is read from the display on the probe. Some pH probes are linked to computers, and the values are logged over time.

TEST YOURSELF

▶ What color would you expect the following solutions to be in the presence of a universal indicator? What pH value would this indicate?

(1) Vinegar

(2) Ammonia solution

(3) Pure water

Neutralization reactions

Substances that do not change the color of simple indicators or turn universal indicator green are described as neutral. Neutral substances have a pH value of 7. Table salt and a solution of sugar and water are neutral substances. Neutral substances are also produced through neutralization reactions. These reactions occur between an acid and an alkali.

HOW DO NEUTRALIZATION REACTIONS WORK?

When an alkali is slowly added to an acid, the hydrogen ions in the acid react with the hydroxide ions in the alkali until they have all reacted. At this point, there are no solitary hydrogen or hydroxide ions present and water is produced. The chemical equation for this reaction is as follows:

Hydrogen ions + Hydroxide ions $\longrightarrow$ Water
$$H^+ + OH^- \longrightarrow H_2O$$

Any other ions from the acid and alkali combine to form a type of compound called a salt.

WHAT IS A SALT?

In chemistry, a salt is a compound that is composed of a metal component and a nonmetal component. Salts are ionic compounds. They contain positively charged ions and negatively charged ions. The charges cancel each other out and the salt has no overall charge.

EXAMPLES OF NEUTRALIZATION REACTIONS

▶ **SODIUM HYDROXIDE AND HYDROCHLORIC ACID**
Sodium hydroxide and hydrochloric acid react as follows:

Sodium hydroxide + Hydrochloric acid $\longrightarrow$
Sodium chloride + Water
$$NaOH_{(aq)} + HCl_{(aq)} \longrightarrow NaCl_{(aq)} + H_2O_{(l)}$$

This produces water and a salt called sodium chloride. Sodium chloride is sometimes called common salt, or table salt, because it is the substance that we put on our food to add flavor or to preserve meat.

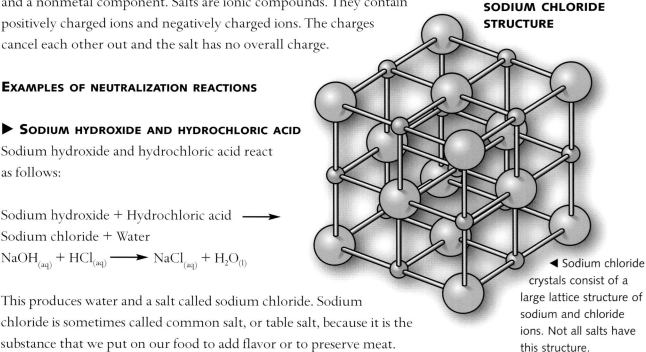

SODIUM CHLORIDE STRUCTURE

◀ Sodium chloride crystals consist of a large lattice structure of sodium and chloride ions. Not all salts have this structure.

However, the sodium chloride that we eat is usually mined or obtained from salt water, rather than created through this type of chemical reaction.

▶ POTASSIUM HYDROXIDE AND SULFURIC ACID

Potassium hydroxide and sulfuric acid also react in a neutralization reaction, illustrated by the equation below.

Potassium hydroxide + Sulfuric acid $\longrightarrow$ Potassium sulfate + Water

$$2KOH_{(aq)} + H_2SO_{4(aq)} \longrightarrow K_2SO_{4(aq)} + 2H_2O_{(l)}$$

Potassium sulfate is a salt, but this is not a salt that we would sprinkle on our food!

USES FOR NEUTRALIZATION REACTIONS AND SALTS

Salts and neutralization reactions have a range of applications in everyday life.

▶ Neutralization is often used in agriculture. When farmers grow crops, the growing plants remove important nutrients from the soil. This can make the soil acidic and unsuitable for further plant growth. Lime is sprayed onto the soil to neutralize the acidity. Lime is also called calcium oxide and is a base.

▶ People suffering from indigestion take tablets called antacids. Indigestion is a condition in which there is too much acid in the stomach. Antacids contain magnesium hydroxide or aluminum oxide, which react with stomach (hydrochloric) acid in the following way:

Magnesium hydroxide + Hydrochloric acid $\longrightarrow$ Magnesium chloride + Water

$$Mg(OH)_{2(aq)} + 2HCl_{(aq)} \longrightarrow MgCl_{2(aq)} + 2H_2O_{(l)}$$

Magnesium chloride is a salt that can pass safely through the body.

DID YOU KNOW?

▶ Bee stings are acidic. Treatments that relieve the pain are alkaline. They work by neutralizing the acid from the sting. Some soaps are alkaline and can relieve bee sting pain.

▶ Wasp stings are alkaline. They are neutralized by applying acidic substances, such as vinegar.

▶ If you are stung by a nettle plant, you will get a painful rash. Nettles contain formic acid. Nettle stings can be alleviated with dock leaves, a weed that often grows near nettles. Dock leaves contain a neutralizing base that relieves the symptoms of the sting.

▲ After a bee stings someone, it tries to fly away. The stinger rips from the bee, causing the bee to die.

TITRATION

Titration is an accurate laboratory technique used to find the exact point at which neutralization occurs. This point is sometimes called the **end point**. There are two parts to titration—making a **standard solution** and carrying out a neutralization reaction.

(1) MAKING A STANDARD SOLUTION

A standard solution is a solution of known concentration. Chemists use a **volumetric flask** to make a standard solution. Volumetric flasks come in a variety of sizes, and each one measures only one volume of solvent. For example, a 250-milliliter volumetric flask cannot be used to measure 100 milliliters of solvent. Volumetric flasks have only one measure mark on them.

The table below shows the steps used to make a standard acid solution:

VOLUMETRIC FLASK

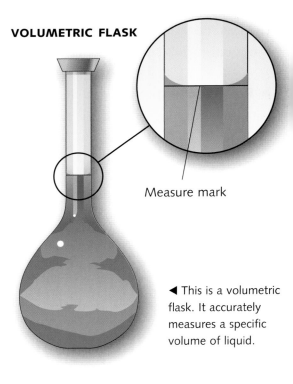

Measure mark

◀ This is a volumetric flask. It accurately measures a specific volume of liquid.

Step	Procedure	Explanation
1	Weigh the solid acid in a **weighing boat**.	The weighing scales are accurate to two or three decimal places and can record the mass accurately.
2	Transfer the solid to a beaker and rinse the weighing boat with distilled water. Put the distilled water from the boat into the beaker. When it is dry, reweigh the boat.	Reweighing the boat confirms that all of the weighed solid has been transferred to the beaker .
3	Add a small amount of distilled water to the beaker. Use a glass rod to crush and dissolve the solid.	The solid is dissolved before it is transferred to the volumetric flask. It is not easy to dissolve a substance in a volumetric flask because of its shape.
4	Once dissolved, pour the solution through a filter funnel into the volumetric flask. Rinse the beaker, funnel, and glass rod with distilled water and add this to the volumetric flask.	All material has been added to the flask and all equipment has been carefully rinsed to make sure that no acid solution remains.
5	Add distilled water until the flask is filled to the measure mark.	When the water level reaches the measure mark, then the flask has been used correctly and the appropriate amount of solvent has been added.
6	Mix the solution before you remove a sample.	Each time the solution is used, it must be mixed well. This action ensures thorough mixing.

(2) CARRYING OUT A NEUTRALIZATION REACTION

A **pipette** and a **burette** are used in this part of the titration process. A pipette accurately measures one volume of solution and is used to add the standard solution to an Erlenmeyer flask. A burette accurately measures different volumes of solution and is used to slowly add the alkali, or acid, to the standard solution.

The table below shows the steps for a neutralization reaction:

▼ This student is filling a pipette with a standard solution of acid that he will put into the Erlenmeyer flask. The burette on the left is filled with an alkali.

Step	Procedure	Explanation
1	Add a specific amount of the standard solution to an Erlenmeyer flask, using a pipette.	The Erlenmeyer flask is used because its shape is ideal for swirling and mixing the solution.
2	Add two or three drops of an indicator to the contents of the Erlenmeyer flask.	The indicator makes the end point of neutralization easier to detect because it will change color when neutralization is reached.
3	Fill a burette with the solution of unknown concentration. This can be either an acid or an alkali. In this example it is an alkali.	The burette must not contain any air bubbles. An initial measurement should be taken. It is important to know the amount of the acid or alkali at the beginning of the experiment.
4	Slowly add the solution from the burette until the indicator just changes color. Constantly swirl the Erlenmeyer flask. Take a reading from the burette.	The flask is swirled to ensure thorough mixing. When the indicator just changes color, another measurement is taken from the burette, so the chemist knows exactly how much solution has been added.
5	Repeat the procedure, but much more slowly, until consistent readings are obtained.	The first reaction was a trial and gives a rough guide to the end point. When repeated, the procedure is slowed down, so the exact point of neutralization can be measured.

▶ This student is adding an alkali to a standard solution. The standard solution contains an indicator. When it changes color, the point of neutralization has been reached.

Measuring pH in a titration

The best indicators change color at a particular pH, rather than gradually over a series of pH values.

(1) A strong acid and a strong alkali

Phenolphthalein, methyl orange, and methyl red are suitable indicators for the titration of a strong acid and a strong alkali. When the alkali is slowly added to the acid and the pH is measured using a pH probe, chemists plot a graph (top right).

(2) A weak acid and a strong alkali

Phenolphthalein is a suitable indicator for this titration. The graph begins at a higher pH because a weak acid is being used. However, the point of neutralization is still distinct, and the remainder of the graph looks much like the first graph.

(3) A strong acid and a weak alkali

Methyl red and methyl orange are suitable indicators for this titration. The starting point of this titration is the same as in the first example because a strong acid is being used. However, because we are using a weak alkali, the point of neutralization is much higher on the vertical part of the graph and the final part of the line is flat.

(4) A weak acid and a weak alkali

This type of titration is rarely carried out because it is very difficult to locate the exact point of neutralization. The change in pH on the line graph is very gradual. The pH of the initial acid and the alkali are both close to 7. The point of neutralization at pH 7 is in the middle of the line on the graph.

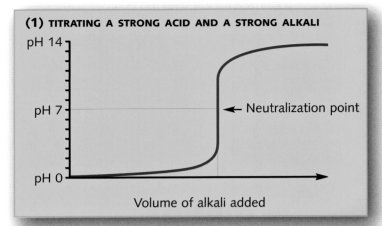

(1) TITRATING A STRONG ACID AND A STRONG ALKALI

pH 14 / pH 7 / pH 0 — Neutralization point — Volume of alkali added

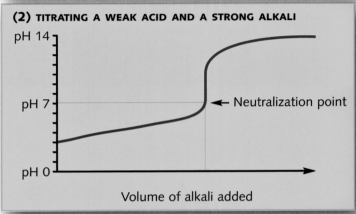

(2) TITRATING A WEAK ACID AND A STRONG ALKALI

pH 14 / pH 7 / pH 0 — Neutralization point — Volume of alkali added

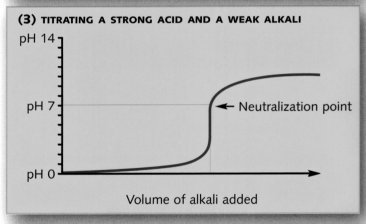

(3) TITRATING A STRONG ACID AND A WEAK ALKALI

pH 14 / pH 7 / pH 0 — Neutralization point — Volume of alkali added

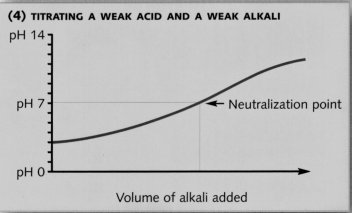

(4) TITRATING A WEAK ACID AND A WEAK ALKALI

pH 14 / pH 7 / pH 0 — Neutralization point — Volume of alkali added

Salts, solutions, and solubility

In chemistry, a solution is a mixture of one or more solutes dissolved in a solvent. The solute is often a solid and the solvent is almost always a liquid. For example, when you add sugar to a cup of tea, the sugar is the solute and the tea is the solvent. Together they make a solution.

Solute + Solvent ⟶ Solution

Some salts are soluble, which means that they will dissolve in a solvent. Other salts are insoluble. Knowing the solubility of salts helps chemists identify them.

SOLVENTS

Water is a universal solvent, which means that it is freely available and can dissolve many substances. There are many other solvents. Nail polish remover is a solvent. It dissolves nail polish, removing it from fingernails. Alcohol is an important solvent for products such as perfume and aftershave. Alcohol dissolves the scent, and when the product is applied to skin, the solvent quickly evaporates and leaves the fragrance behind.

SOLUBILITY

Solubility is a measure of how readily a solute will dissolve in a solvent. Have you ever tried to dissolve sugar in cold water or milk? You may have noticed if you sprinkle sugar onto cereal, there is often sugar left at the bottom of the bowl. However, when sugar is added to a hot liquid, it will dissolve with minimal stirring. This indicates that substances dissolve more easily in hot solvents than in cold solvents. The opposite is true if the solute is a gas.

Solubility is usually dependent on temperature. Solubility is measured by the number of grams of solute that will dissolve in 100 grams of solvent at a particular temperature. There is no pattern to the solubility of different solutes, and it is not easy to predict. The following graph shows the solubility of two common salts.

SOLUBILITY OF TWO SALTS WITH TEMPERATURE

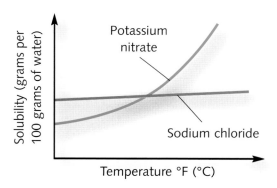

SATURATED SOLUTIONS

Saturated solutions are solutions in which no more solute will dissolve at a given temperature. For example, at room temperature, 30 grams of sodium chloride will easily dissolve into 100 milliliters (100 grams) of water. If an extra 10 grams of sodium chloride is added, 4 grams remain undissolved because the solution has become saturated. Therefore, the solubility of sodium chloride is 36 grams of solute in 100 grams of water at 68°F (20°C).

SUMMARY

▶ 30 grams sodium chloride + 100 grams water = 130 grams solution

▶ 130 grams solution + 10 grams sodium chloride = 140 grams in the container

▶ But 4 of the 40 grams of added sodium chloride remain undissolved:
40 grams − 4 grams = 36 grams

▶ In total, 36 grams of sodium chloride will dissolve in 100 grams of water.

SOLUBILITY RULES

If a substance can dissolve in water, it is soluble, and it will dissolve to a certain concentration. However, once a solution becomes saturated, the substance is no longer soluble. If a substance is described as insoluble this means that the majority of it will not dissolve into solution. When an insoluble substance is added to a solvent, it forms a **suspension**. A suspension is a solid substance that floats in a solvent.

The solubility of salts follows specific rules:

Type of salt	Soluble	Insoluble
Chlorides	Most are soluble	Silver chloride and lead chloride
Sulfates	Most are soluble	Barium, calcium, and lead sulfates
Nitrates	All are soluble	None are insoluble
Carbonates	Sodium and potassium carbonates	Most are insoluble
Acetates	All are soluble	None are insoluble
Sodium, potassium, and ammonium salts	All are soluble	None are insoluble

▲ These stalactites and stalagmites are made of calcium carbonate, which is insoluble in water. Calcium carbonate reacts with rainwater and forms soluble calcium bicarbonate. When the water drips from cave ceilings, calcium carbonate **precipitates** from the water droplets and creates these incredible natural structures.

TIME TRAVEL: CARBON SINKS

Carbon dioxide is an acidic gas. This means that it dissolves in water to form a weakly acidic solution called carbonic acid. This is a very important process because it helps control climate change. Carbon dioxide is a greenhouse gas, and when it accumulates in the atmosphere, it contributes to global warming. Approximately two-thirds of the earth is covered in water, and much of the carbon dioxide produced through human activity dissolves in the oceans. This is called the carbon sink.

Throughout history, massive amounts of carbon dioxide released by animals and other natural processes have been absorbed into the carbon sink, and the amount of carbon dioxide in the atmosphere has remained steady for the last 650,000 years. However, since the development of electricity, cars, and factories, carbon dioxide emissions have significantly increased.

During the 1980s and 1990s, about half of all carbon dioxide emissions remained in the atmosphere—20 percent was absorbed by plants through photosynthesis and 30 percent went into the oceans as carbon sink.

Today, the amount of carbonic acid in the oceans has increased dramatically, changing the chemical composition of the water. The increased acidity has a direct effect on organisms, such as coral, whose bodies are made of carbonates. A chemical reaction occurs between the seawater and the carbonates in the coral. The coral decomposes and disappears. This is another reason why it is important to reduce carbon emissions.

▼ When excess carbon dioxide is dissolved in the oceans, the seawater becomes more acidic. This can harm marine organisms.

MAKING PREDICTIONS ABOUT SOLUBILITY

A number of rules can be applied to predict the solubility of combined substances. For example, the reaction between silver nitrate and sodium chloride produces silver chloride and sodium nitrate. The equation for this reaction is as follows:

Silver nitrate + Sodium chloride $\longrightarrow$ Silver chloride + Sodium nitrate

$AgNO_3 + NaCl \longrightarrow AgCl + NaNO_3$

Using the rules from the solubility table on page 20, we know that silver nitrate, sodium chloride, and sodium nitrate are soluble. In an aqueous solution, silver chloride is insoluble, so it would appear as a precipitate—a solid substance formed when two or more solutions react. Symbols representing the state of the substances—solid, liquid, gas, or aqueous—in this reaction can be inserted into the equation:

$AgNO_{3(aq)} + NaCl_{(aq)} \longrightarrow AgCl_{(s)} + NaNO_{3(aq)}$

TEST YOURSELF

▶ What would you see when ammonium chloride and silver nitrate are combined? Write a balanced chemical equation with the state symbols for this reaction.

SEPARATING SOLUTIONS

There are a wide range of chemical techniques that will separate solutions.

EVAPORATION

When a solution is gently heated, the liquid evaporates from the surface until only the solid remains. A dish with a large surface area should be used for evaporation; this is called an evaporating dish. The dish is placed over a water bath because once the solution evaporates, some of the remaining solid material may break down and the solution can sputter dangerously. It is best to evaporate most of the solution over the hot water bath and leave the rest to evaporate near a sunny window.

▲ This copper sulfate solution is evaporating over a water bath.

CRYSTALLIZATION

This technique is similar to evaporation—a solution is heated and the solvent begins to evaporate. This concentrates the amount of solute in the solution. When the solution is saturated, it is removed from the heat and left to cool. Crystals of solute will appear.

CHROMATOGRAPHY

In this technique, a drop of solution is placed on a surface, such as absorbent filter paper. One end of the paper is set in a beaker of solvent. The solvent should slowly travel up the paper and carry the solution with it. When the paper is removed and dried, small spots of solute are evident. The solute has successfully separated from its solution and can be scraped off the paper.

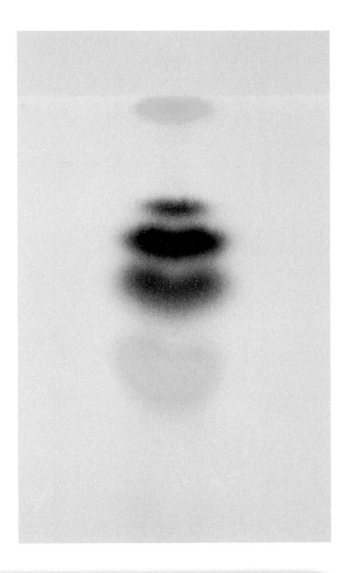

▶ This is a chromatogram of an extract from meadow grass. The different colored bands represent different chemicals. The orange band (top) is carotene, the green bands are chlorophyll, and the yellow (bottom) represents other carotenes.

TIME TRAVEL: INTO THE FUTURE

Scientists are trying to find out why the tumors of some cancer patients grow faster than others. They hope that a greater understanding of a tumor's behavior may reduce cancer deaths in the future. Chemistry is providing some answers. Fried foods contain linoleic and linolenic acids. Research has shown that the proportions of these two acids in food can affect how quickly cancerous tumors grow. In one study, when cancer patients ate food that contained these acids in a ratio of 9:1 or higher, their tumors grew more quickly.

A team of students at a school in Kansas tested french fries in a selection of local restaurants.

Their experiment involved (1) extracting oil from the french fries using a solvent; (2) separating the particles in the oil using a centrifuge (a rapidly spinning machine); (3) evaporating the solvent; and (4) reacting another solvent with the fatty acids to find out the ratios of linoleic to linolenic acids.

They found that some french fries contained linoleic and linolenic acids at an incredibly high ratio of 17:1, whereas some had a much lower ratio of 3:1, and these had a double layer of batter on them. Scientists are trying to determine whether the low-ratio french fries are better for us, and if so, whether it is the batter that makes them healthier.

USEFUL SALTS

Salts have been used by people for many years. Here, we will explore some of the uses for salts in the past, present, and the future.

ANTACIDS

Antacids are taken by people who suffer from heartburn. Some antacids contain the base sodium hydrogencarbonate, or sodium bicarbonate. The stomach naturally produces hydrochloric acid to help digest food. However, if someone eats a particularly rich or large meal, the stomach may produce excess acid. Unfortunately, this can cause a lot of discomfort. The antacids react with the stomach's hydrochloric acid in the following way:

Sodium bicarbonate + Hydrochloric acid $\longrightarrow$
Sodium chloride + Carbon dioxide + Water
$$NaHCO_{3(aq)} + HCl_{(aq)} \longrightarrow$$
$$NaCl_{(aq)} + CO_{2(g)} + H_2O_{(l)}$$

The sodium bicarbonate neutralizes the acid. Because of the buildup of carbon dioxide gas, you may feel like "burping" afterward.

GLAUBER'S SALT

The common name for this salt is sodium sulfate, and it is a **hydrated salt**. This means that water molecules are present in the salt structure. This salt was discovered by Johann Glauber in the 1600s and was originally used as a laxative. Today, the main use for Glauber's salt is as an ingredient in powdered laundry detergents.

Other uses for sodium sulfate are:

▶ to prevent the formation of small air bubbles in molten glass;

▲ When sodium bicarbonate reacts with acid, it releases bubbles of carbon dioxide.

▶ to help dyes stick to materials during textile manufacture;

▶ as a heat storage component in solar heating systems. This is being researched for use in the future.

Approximately half of the world's supply of Glauber's salt is found naturally, particularly in lakes in Canada. The other half is synthesized—some is produced as a by-product of other chemical reactions. For example, the production of both hydrochloric and sulfuric acids results in the formation of sodium sulfate.

Sodium chloride + Sulfuric acid $\longrightarrow$
Sodium sulfate + Hydrogen chloride
$$2NaCl_{(s)} + H_2SO_{4(aq)} \longrightarrow Na_2SO_{4(aq)} + 2HCl_{(g)}$$

GYPSUM

Gypsum's chemical name is calcium sulfate, and its formula is $CaSO_4·2H_2O$. Gypsum is used in chalk and cement. One of its main uses is in drywall. Drywall is a common material for the construction of interior walls and ceilings. It is made by forming gypsum into flat sheets, which are then sandwiched between two pieces of heavy paper. Gypsum is a hydrated compound. When it is exposed to heat, such as in a house fire, the water in its structure is converted to steam. The gypsum cannot become hotter or burn until this water has completely evaporated. Therefore, drywall is fire-resistant.

Gypsum occurs naturally and can also be manufactured. When fossil fuels are burned, sulfur is released as sulfur dioxide. This gas is passed through special chimneys called "scrubbers" where the sulfur dioxide reacts with limestone. This "cleans" the gas and produces calcium sulfate as a by-product. The gypsum produced by this method is very pure. Gypsum can also be produced as a by-product of phosphate fertilizer refinement.

▲ Glauber's salt (sodium sulfate) is found naturally in lake beds in Canada.

▶ The inside walls of this house were constructed with drywall that is made from gypsum.

5

Making salts

The five main ways to make salts are (1) to react a metal and an acid; (2) to react an insoluble base and an acid; (3) to react an alkali and an acid; (4) to react a carbonate and an acid; and (5) a precipitation reaction between two salts.

MAKING SALTS USING METALS

Acids react with most metals and the reactions always produce salts. These reactions also produce hydrogen gas. The general equation for this reaction is as follows:

Metal + Acid ⟶ Salt + Hydrogen

This reaction is vigorous if a highly reactive metal, such as sodium, is used.

POTASSIUM

Potassium is a highly reactive metal and will cause a violent reaction with acids such as hydrochloric or sulfuric acid. It is also very reactive with water, which is described by chemists as an extremely weak acid.

Potassium + Hydrochloric acid ⟶ Potassium chloride + Hydrogen

$$2K_{(s)} + 2HCl_{(aq)} \longrightarrow 2KCl_{(aq)} + H_{2(g)}$$

ZINC

Zinc is not as reactive as potassium. When a less reactive metal combines with an acid, a salt and hydrogen gas are still produced, but the reaction will be less vigorous.

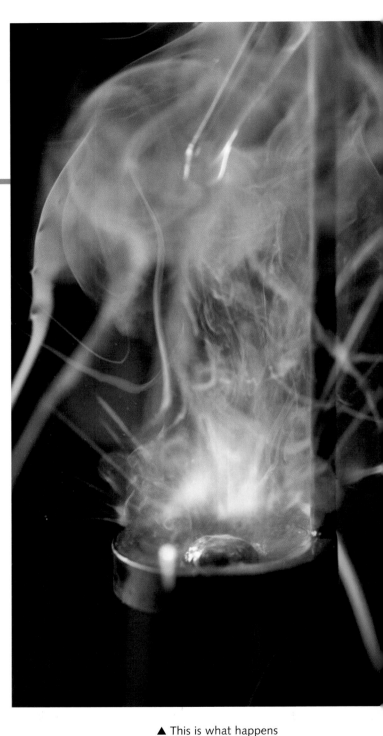

▲ This is what happens when water is dropped onto potassium metal. This reaction produces a lot of heat, potassium hydroxide, and hydrogen gas.

Zinc + Hydrochloric acid $\longrightarrow$
Zinc chloride + Hydrogen

$$Zn_{(s)} + 2HCl_{(aq)} \longrightarrow ZnCl_{2(aq)} + H_{2(g)}$$

TIN

Other metals that are low in the reactivity series will only produce a salt over a long period of time. They may not appear to react at all. For example, tin will occasionally produce a bubble of hydrogen gas when reacted with acid.

Tin + Hydrochloric acid $\longrightarrow$
Tin chloride + Hydrogen

$$Sn_{(s)} + 2HCl_{(aq)} \longrightarrow SnCl_{2(aq)} + H_{2(g)}$$

▲ Zinc reacts with hydrochloric acid and produces zinc chloride and bubbles of hydrogen gas.

TEST YOURSELF

▶ Write balanced chemical equations for each of the following chemical changes:

(1) Calcium and sulfuric acid

(2) Copper and diluted nitric acid

(3) Magnesium and hydrochloric acid

USING ACIDS TO MAKE SALTS

It is possible to produce a wide range of salts by varying not only the type of metal, but also the type of acid used in a reaction. The table below summarizes the salts that are formed with different acids.

Type of acid	Type of salt produced
Hydrochloric acid	Chloride
Sulfuric acid	Sulfate
Nitric acid	Nitrate
Phosphoric acid	Phosphate

OBTAINING SOLID SALT CRYSTALS FROM SOLUTION

Each salt in the above table is soluble and can be found in a mixture of unreacted acid and water. The following method is used to obtain pure salt crystals:

(1) Make sure the reaction is complete. To do this, add metal to the reaction mixture until the bubbles of hydrogen gas are no longer visible.

(2) Heat the solution in a water bath until crystals begin to form.

(3) Leave the solution in an evaporating dish near a sunny window until the crystals of salt are fully formed.

INVESTIGATE

▶ Place an iron nail in an acid solution, such as water mixed with vinegar, lemon juice, or soda pop. Observe the nail after a few days. What has happened to it?

MAKING SALTS USING
INSOLUBLE BASES

A salt can be made from the reaction between an acid and an insoluble base. This reaction also produces water. The general equation for this reaction is as follows:

Acid + Insoluble base ⟶ Salt + Water

Oxides are usually insoluble with the exception of ammonium oxide and the group I oxides (sodium oxide, potassium oxide, and lithium oxide). Zinc oxide, copper oxide, and magnesium oxide are insoluble bases.

The table below summarizes the salts that are made from two common insoluble bases:

Acid	Insoluble base	Soluble salt
Sulfuric acid	Iron oxide	Iron sulfate
Nitric acid	Zinc oxide	Zinc nitrate

COPPER OXIDE + HYDROCHLORIC ACID

Copper oxide is an insoluble base. It will react with hydrochloric acid in the following way:

Copper oxide + Hydrochloric acid ⟶
Copper chloride + Water

$$CuO_{(s)} + 2HCl_{(aq)} \longrightarrow CuCl_{2(aq)} + H_2O_{(l)}$$

HOW TO CARRY OUT THE REACTION

The procedure described below is used to react an acid and an insoluble base, producing a soluble salt.

(1) Add a known amount of hydrochloric acid to a boiling tube and warm it. The rise in temperature accelerates the reaction.

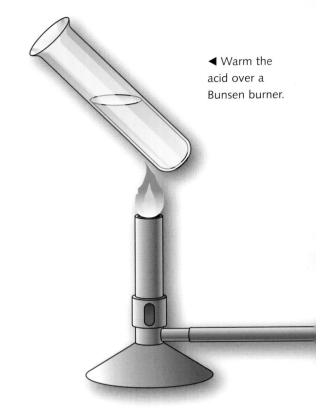

◀ Warm the acid over a Bunsen burner.

(2) Add solid copper oxide to the warmed acid, one spatula at a time. Continue to do this until the copper oxide no longer "disappears" into the solution. As the copper oxide is added, the reaction proceeds, and the solution begins to turn greenish-blue as copper chloride forms. An excess of copper oxide is added to ensure that all the hydrochloric acid has reacted.

(3) **Filter** the excess copper oxide from the solution, which now contains copper chloride and water.

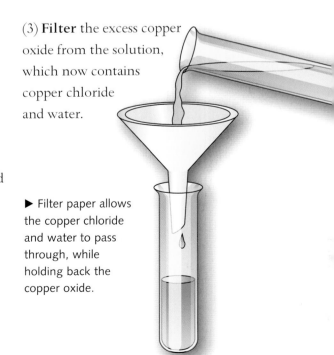

▶ Filter paper allows the copper chloride and water to pass through, while holding back the copper oxide.

(4) Pour the copper chloride solution into an evaporating dish and indirectly heat it above a large beaker of water. This limits the sputtering that may occur and preserves the quality of the final crystals.

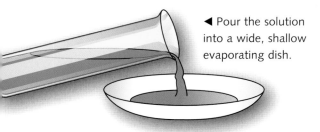

◀ Pour the solution into a wide, shallow evaporating dish.

▼ Place the evaporating dish over a water bath, and heat with a Bunsen burner.

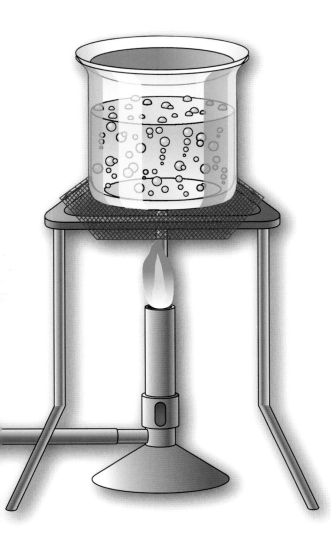

(5) When the first crystals appear, remove the evaporating dish from the water bath and place near a sunny window. The crystals will appear naturally as the water slowly evaporates. The slow evaporation encourages large formations of crystal. If the solution continued to heat over a water bath, the crystals would form more rapidly but would be much smaller.

▲ Large crystals of copper chloride will form gradually as the remaining water evaporates.

SUMMARY

The steps involved in making a salt from an acid and an insoluble base are summarized by the following terms:

▶ Mixing
▶ Filtration
▶ Evaporation
▶ Crystallization

TEST YOURSELF

▶ How would you make pure crystals of magnesium sulfate from an insoluble base and an acid? Give the names of the reactants you would use and the steps for making the crystals.

▶ Why would the answer from the question above not be a good method for making a pure sample of potassium sulfate crystals?

MAKING SALTS USING ALKALIS

A soluble salt can be made from the reaction between an acid and an alkali. Alkalis are soluble bases and usually contain hydroxide ions. Ammonia solution is also an alkali. The general equation for this reaction is as follows:

Acid + Alkali ⟶ Salt + Water

SODIUM HYDROXIDE + HYDROCHLORIC ACID

Sodium chloride can be made by reacting sodium metal and chlorine gas, but this reaction is extremely vigorous and dangerous. A safer way to produce sodium chloride crystals is to react an acid with an alkali.

Sodium hydroxide + Hydrochloric acid ⟶ Sodium chloride + Water

In this case, all the reactants and the final salt are soluble. Therefore, retrieving the final product requires careful laboratory methods.

HOW TO CARRY OUT THE REACTION

The following steps are used in the laboratory to react sodium hydroxide with hydrochloric acid.

(1) Measure 25 milliliters of sodium hydroxide into an Erlenmeyer flask.

(2) Add three or four drops of phenolphthalein indicator. This indicator is pink in alkaline solutions. It is important to use a known amount of sodium hydroxide because the experiment will be repeated later without the indicator.

(3) Fill a burette with hydrochloric acid. Record the volume of acid at the start of the experiment.

(4) Add the hydrochloric acid slowly to the sodium hydroxide in the Erlenmeyer flask. At the same time, gently swirl the flask to mix the acid and base.

(5) The indicator will turn colorless when all of the alkali has been neutralized by the acid. At this point, do not add any more acid.

▼ This Erlenmeyer flask contains sodium hydroxide and phenolphthalein, which is pink in basic solutions. Hydrochloric acid will be added from the burette.

(6) Take a final reading from the burette. To calculate the exact amount of acid added, subtract the second reading from the first. The purpose of this part of the procedure is to measure how much acid is needed to neutralize the initial volume of alkali.

(7) Repeat the experiment without an indicator. The indicator is not necessary because the amount of acid that is required for neutralization has been determined. The addition of an indicator would make the final product impure.

(8) The solution remaining in the Erlenmeyer flask contains sodium chloride and water. Put this in an evaporating dish and heat it indirectly above a beaker of water.

(9) When crystals appear, place the evaporating dish near a sunny window. The crystals will slowly form.

TEST YOURSELF

▶ Explain how you would make pure crystals of potassium chloride using potassium hydroxide and hydrochloric acid. State the method that you would use.

DID YOU KNOW?

▶ Forensic scientists use phenolphthalein to test for the presence of blood at a crime scene. They wipe a cotton swab over the suspected sample of blood. Next they add a few drops of alcohol, followed by a few drops of phenolphthalein indicator. Finally, they add a few drops of hydrogen peroxide.

If the sample on the cotton swab turns pink, then blood is present. Once a scientist has confirmed the presence of blood, they can find out the blood type and analyze the DNA. They may be able to connect this information to a person.

The formation of the blood drop on the ground at the crime scene can also tell the scientist whether the person was standing still or moving, and in which direction they were moving.

▶ **This forensic scientist is cutting a sample of material from blood-stained jeans. He will analyze the DNA and try to link a suspect with the victim and the crime scene.**

MAKING SALTS USING CARBONATES

Another method of making a salt is the reaction between a metal carbonate and an acid. Most metal carbonates are insoluble and the general equation for this reaction is as follows:

Metal carbonate$_{(s)}$ + Acid$_{(aq)}$ ⟶
Metal salt$_{(aq)}$ + Carbon dioxide$_{(g)}$ + Water$_{(l)}$

Once all of the metal carbonate has been reacted with the acid, the soluble salt will be present in the remaining solution. However, as with the previous examples, to ensure that all of the acid has reacted, it is important to use an excess of metal carbonate. This is achieved by adding so much solid that the bubbling action—indicating a reaction—stops, even though there is still solid carbonate remaining in the solution.

MAGNESIUM CARBONATE + SULFURIC ACID

Magnesium carbonate will react with sulfuric acid in the following way:

Magnesium carbonate + Sulfuric acid ⟶
Magnesium sulfate + Carbon dioxide + Water
$MgCO_{3(s)} + H_2SO_{4(aq)} \longrightarrow MgSO_{4(aq)} + CO_{2(g)} + H_2O_{(l)}$

HOW TO CARRY OUT THE REACTION

(1) Add a known volume of acid to a test tube.

(2) Spoon the solid metal carbonate into the acid, one spatula at a time. Do this until bubbles are no longer produced and no more metal carbonate will dissolve.

▶ This photograph shows magnesium carbonate reacting with acid, as in step 2 of the experiment. The reaction is still proceeding because you can see carbon dioxide bubbles forming.

(3) Filter the mixture to remove unreacted metal carbonate from the salt solution. No unreacted acid should remain in the solution.

(4) Place the salt solution in an evaporating dish on a water bath and heat it.

(5) When crystals begin to appear, the evaporating dish contains a saturated solution. Leave the apparatus near a sunny window until crystals of magnesium sulfate have formed.

WEATHERING

Calcium carbonate (limestone) buildings can be corroded by acid rain, which is a dilute solution of sulfuric acid. This produces calcium sulfate, carbon dioxide, and water. Limestone is particularly at risk because it is very porous. This means that the rain easily soaks into the rock's tiny pores and corrodes it. The salt produced by this reaction, calcium sulfate, is insoluble, so the rock does not simply disintegrate. However, the two other products—water and carbon dioxide—do leave the rock. The water evaporates and the carbon dioxide passes into the atmosphere. The limestone is gradually worn away. The Statue of Liberty in New York City had to be restored because of damage caused by acid rain. The statue is made from copper. Rain has reacted with the copper, which is why it looks green.

▶ Acid rain is a weak acid. It corrodes buildings that are made from limestone, such as the Louisiana State Capitol building and thousands of other buildings around the world.

TEST YOURSELF

▶ Marble is also made from calcium carbonate, but it does not become corroded like limestone. Use the Internet or the library to find out why differences between marble and limestone mean that marble is not easily corroded.

INVESTIGATE

▶ **Ask an adult before trying this experiment**.

You will need citric acid crystals (available at pharmacies), baking soda, and powdered sugar. Mix the ingredients in equal proportions and put a small amount of the mixture in your mouth. What happens? Use library books or the Internet to help you write the word equation for this chemical change.

MAKING SALTS USING PRECIPITATION REACTIONS

Salts can be made through a precipitation reaction. This method is used to produce an insoluble salt. A precipitate is a solid produced from the chemical reaction of two or more solutions. This method is sometimes called **double decomposition**. A general equation for the reaction is as follows:

Soluble salt 1 + Soluble salt 2 $\longrightarrow$ Insoluble salt + Soluble salt 3

LEAD NITRATE + SODIUM SULFATE

Lead sulfate is insoluble and can be made from the soluble salts lead nitrate and sodium sulfate. The equation for this reaction is written below.

Lead nitrate + Sodium sulfate $\longrightarrow$ Lead sulfate + Sodium nitrate

$$Pb(NO_3)_{2(aq)} + Na_2SO_{4(aq)} \longrightarrow PbSO_{4(s)} + 2NaNO_{3(aq)}$$

From this equation, it is clear where the name "double decomposition" comes from—the metal ions exchange places.

HOW TO CARRY OUT DOUBLE DECOMPOSITION

The following steps are used to obtain a pure and dry sample of an insoluble salt.

(1) Mix the two soluble salts in equal proportions. This produces an insoluble precipitate.

(2) Filter the insoluble precipitate from the remaining mixture of solutions. Keep the solid residue because it contains the salt. Dispose of the filtrate (liquid).

(3) If the precipitate has a very thick consistency it may need to be filtered using vacuum pressure. This can be done with a Buchner funnel and Buchner flask. The apparatus is attached to a water tap, which is then turned on full force. The water rushing through the apparatus creates pressure that is sufficient to pull liquid through the filter funnel and into the flask—like a vacuum cleaner's suction. This equipment can remain connected to allow the residue to continue drying.

(4) The residue on the filter paper is the insoluble salt. Lay the paper flat and allow the salt residue to completely dry in a dust-free environment.

BUCHNER FILTRATION

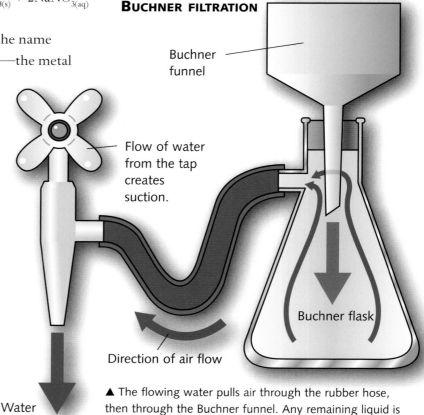

Buchner funnel

Flow of water from the tap creates suction.

Direction of air flow

Buchner flask

Water

▲ The flowing water pulls air through the rubber hose, then through the Buchner funnel. Any remaining liquid is suctioned through the Buchner funnel and into the flask.

Making paint

Precipitation reactions are used to produce colored **pigments** that are added to paints. The table below shows the colors of different insoluble precipitates:

Name of precipitate	Color of precipitate
Lead carbonate	White
Lead chromate	Yellow
Iron (III) ferrocyanide	Blue
Copper carbonate	Green

▼ This bright yellow precipitate is lead chromate. It has been formed by dripping potassium chromate into a lead (ll) nitrate solution.

INVESTIGATE

▶ Ask an adult before trying this experiment.

In the sixteenth and seventeenth centuries, Mogul artists in India made their paints from natural materials, such as stones and minerals, which they ground into fine powders. After grinding, the material was washed to remove impurities. The insoluble material produced was called a pigment, and it was mixed with a binding agent. The main binding agent was gum from acacia trees. To allow the pigment to flow better and stick to the cloth, the artists added honey to it.

Use the Internet to research Mogul techniques; try and copy their methods to make your own paint out of turmeric or saffron. Saffron is expensive, so ask an adult's permission before using it. You may also be able to extract pigments from food, such as red cabbage or blackberries.

▲ This Mogul painting was created in 1633.

DID YOU KNOW?

▶ Stink bombs contain a solution of ammonium sulfide $(NH_4)_2S$. When you crush them, the solution diffuses into the air creating a pungent, choking smell. Chemistry can be used to deactivate a stink bomb. If the bomb is opened while it is submerged in a solution of lead nitrate, the following chemical reaction occurs.

Ammonium sulfide + Lead nitrate ⟶ Lead sulfide + Ammonium nitrate

Lead sulfide is a brown, insoluble precipitate. And most importantly, it has no smell!

6

Uses and manufacture of acids

Acids are essential to industry and have been in use for hundreds of years. Hydrochloric acid is used to make PVC plastic, which is used for pipes and numerous pharmaceutical products. Sulfuric acid is used to make fertilizers and detergents. Nitric acid is used to make explosives and rocket fuel. However, before acids can be utilized, they must be manufactured.

HYDROCHLORIC ACID

Hydrochloric acid (HCl) is a solution of hydrogen chloride gas dissolved in water. Hydrogen chloride is a white, soluble gas that dissolves readily in water. Hydrochloric acid is a strong acid. It gives away its H^+ ions easily.

USES OF HYDROCHLORIC ACID

▶ Manufacture of other chemicals
The main use of hydrochloric acid is for the production of other chemicals, such as calcium chloride (used in road salt), nickel chloride (used for electroplating), and zinc chloride (used in batteries).

▶ Water softening
Ion exchange resins are used to soften water, which is important in the treatment of drinking water. Hydrochloric acid is used in the manufacture of ion exchangers. The exchangers swap the ions that make the water hard for other ions. For example, calcium and sodium ions are found in hard water. When the water passes through an ion exchanger, these ions are removed.

▶ Control of pH
Hydrochloric acid is used to regulate pH conditions. For example, hydrochloric acid helps maintain the carefully controlled water pH that is required by the food and pharmaceutical industries. It is also used to maintain the pH of swimming pools.

▶ Steel pickling
Acid removes rust from iron or steel before the metal is made into cans or wires.

◀ A mixture of salt and sand is deposited onto the road surface from the back of this snowplow. Calcium chloride is the most effective salt to use for melting ice and snow. It also prevents ice and snow from bonding with the surface of the road.

TIME TRAVEL: HYDROCHLORIC ACID

Hydrochloric acid was first discovered in A.D. 800 by Jabir ibn Hayyan, who was born in Iran, but spent much of his life in Iraq and Yemen. He is also credited with the discovery of many other chemicals including citric acid, nitric acid, and acetic acid.

Medieval alchemists used hydrochloric acid in their search for "the philosopher's stone," which was believed to hold the secret of eternal life. Alchemists made hydrochloric acid by reacting sodium chloride with sulfuric acid. The equation for this reaction is as follows:

Sodium chloride + Sulfuric acid $\longrightarrow$
Sodium sulfate + Hydrogen chloride
$2NaCl_{(s)} + H_2SO_{4(aq)} \longrightarrow Na_2SO_{4(aq)} + 2HCl_{(g)}$

Next, the hydrogen chloride would have been dissolved in water. This was done very carefully because of the extreme solubility of this gas.

▼ **Jabir ibn Hayyan, sometimes referred to as the Father of Chemistry, is shown here teaching at a school in Mesopotamia.**

During the United Kingdom's Industrial Revolution in the 1800s, demand increased for alkaline substances, such as soda ash (sodium carbonate) and sodium and potassium hydroxides, and large scale production of these materials was developed. One of the by-products of the processes was hydrogen chloride gas and until 1863, the gas was released directly into the air. In 1863, the "Alkali Act" was passed in Parliament, which required manufacturers to dissolve the waste gas in water to reduce air pollution. This produced large amounts of hydrochloric acid.

In the 1890s, production of industrial alkalis was replaced by the Solvay process. This did not produce a hydrochloric acid by-product, so other means of producing the acid were sought. In 1926, hydrochloric acid also became important in the manufacture of PVC (polyvinyl chloride). This plastic has changed our world. Today, approximately 22 million tons (20 million t) of hydrogen chloride gas is produced annually and used to make hydrochloric acid.

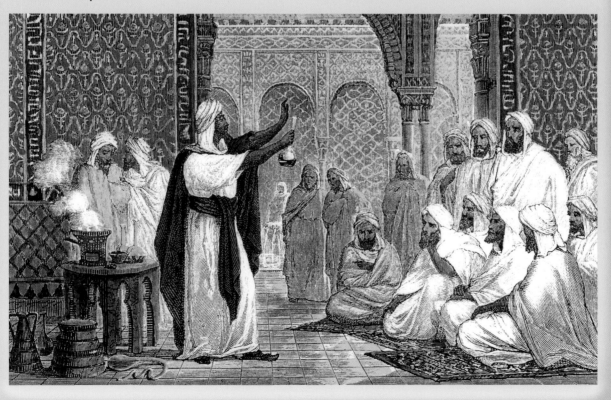

HOW IS HYDROCHLORIC ACID PRODUCED TODAY?

▶ Hydrochloric acid is produced as a by-product of the manufacture of chlorine and an alkaline salt, such as sodium hydroxide. In this process, electricity is passed through a sodium chloride solution. This produces three substances—hydrogen, chlorine, and sodium hydroxide. When the hydrogen and chlorine recombine, hydrogen chloride gas is produced.

$$Cl_{2(g)} + H_{2(g)} \longrightarrow 2HCl_{(g)}$$

The hydrogen chloride gas is dissolved in water to produce hydrochloric acid.

▶ A reaction between sulfuric acid and sodium chloride also produces hydrochloric acid.

▶ Hydrochloric acid is a part of the production of **organic compounds**.

DID YOU KNOW?

▶ Phosgene ($COCl_2$) is a chemical weapon that was used during World War I. When phosgene dissolves deep in the lungs, it forms hydrochloric acid. This acid damages the lining of the lungs and causes them to fill with fluid. A similar process occurs with mustard gas. When it dissolves in moist areas of the body, such as lungs or eyes, hydrochloric acid forms, causing irreversible tissue damage.

▶ In 1938, an accident led to the discovery of Teflon. The scientists involved believed that they could make a new refrigerant by mixing a compound called TFE with hydrochloric acid. They filled canisters with TFE and froze them. The next day they opened the canisters and tried to pour the TFE into vessels of hydrochloric acid. The contents of the first canister would not pour out and felt heavier than expected. The canister was opened, and inside was a waxy white solid. A chemical reaction had taken place. This new chemical proved to be a good lubricant and was chemically unreactive; it also had a very high melting temperature. This solid became known as Teflon and is now used for the surfaces of nonstick pans.

Organic compounds include Teflon, PVC, and CFCs. CFCs (chlorofluorocarbons) were used in refrigerators and aerosols until scientists discovered that they harmed the earth's ozone layer.

SULFURIC ACID

Sulfuric acid (H_2SO_4) was discovered around 1,200 years ago by alchemists and has been a useful chemical ever since.

THE CHEMISTRY OF SULFURIC ACID

Sulfuric acid has some unique properties. It can be used to extract water from many chemicals. This makes it useful as a drying agent—it can remove moisture from air and from compounds such as sugar and starch. Sulfuric acid is also an **oxidizing agent**. In this reaction it can dissolve some metals and form sulfur dioxide gas.

▲ Sulfuric acid poured into a beaker of sugar causes this drying reaction. The black pillar is dried sugar, or carbon.

USES FOR SULFURIC ACID

▶ Approximately 60 percent of all manufactured sulfuric acid is used to make phosphoric acid. Phosphoric acid is important for the production of fertilizers and detergents. Another fertilizer, ammonium sulfate, is also made using sulfuric acid.

Ammonia + Sulfuric acid ⟶ Ammonium sulfate

▶ Sulfuric acid is used to make paint, soaps, dyes, plastics, and soapless cleansers, such as shower gel. It is also used in the manufacture of toilet paper.

▶ Sulfuric acid is an ingredient in the production of aluminum sulfate, which is important in the papermaking process. Aluminum sulfate is produced from the reaction between sulfuric acid and aluminum oxide.

▲ Sulfuric acid is involved in the manufacture of paints.

Use of sulfuric acid	Percentage of acid
Fertilizers	30
Paints and pigments	14
Detergents and soaps	13
Plastics	13
Dyes	2
Steelmaking	1
Other	27

SELF-CLEANING GLASS

One of the more interesting uses for sulfuric acid is in the manufacture of "self-cleaning" glass. This glass can break down a layer of grease up to 200 nanometers thick per day!

The self-cleaning glass is covered with a clear film of titanium dioxide. This top layer is a **catalyst**. It accelerates chemical reactions in the presence of the sun's energy. When sunlight shines onto the titanium dioxide, the energy from the sun makes the chemical layer reactive. As dirt and grease accumulate on the layer, a chemical reaction takes place. The products of the reaction are usually carbon dioxide and water, which run off the windows, leaving them clean.

▲ In the future, self-cleaning glass could be used for high-rise buildings. At the moment, people must scale these buildings to clean the glass.

MANUFACTURE OF SULFURIC ACID: THE CONTACT PROCESS

In the contact process, sulfur dioxide is turned into sulfur trioxide, which is then dissolved in water to produce sulfuric acid. The following steps describe the contact process.

(1) The production of sulfur dioxide begins when sulfur is burned in air.

Sulfur + Oxygen $\longrightarrow$ Sulfur dioxide

$$S_{(l)} + O_{2(g)} \longrightarrow SO_{2(g)}$$

Notice that liquid sulfur is used. This is sprayed into a furnace.

(2) In the production of sulfur trioxide, sulfur dioxide is mixed with air and passed through a catalyst called vanadium pentoxide. The catalyst accelerates the reaction. The mixture is heated to 842°F (450°C) and sulfur trioxide is produced.

Sulfur dioxide + Oxygen $\longrightarrow$ Sulfur trioxide

$$2SO_{2(g)} + O_{2(g)} \longrightarrow 2SO_{3(g)}$$

(3) A very concentrated form of sulfuric acid, called oleum, is produced when the sulfur trioxide is passed through an absorber that contains sulfuric acid. Sulfur trioxide dissolves in the acid and produces oleum.

Sulfur trioxide + Sulfuric acid $\longrightarrow$ Oleum

$$SO_{3(g)} + H_2SO_{4(aq)} \longrightarrow H_2S_2O_{7(l)}$$

(4) Oleum is carefully dissolved in water to produce sulfuric acid (H_2SO_4).

Oleum + Water $\longrightarrow$ Sulfuric acid

$$H_2S_2O_{7(l)} + H_2O_{(l)} \longrightarrow 2H_2SO_{4(aq)}$$

When highly concentrated solutions are produced, the state symbol "l" is used instead of "aq" because very little water is present. Many of the reactions involved in the contact process release heat—they are exothermic reactions. This heat is not wasted. Instead, the heat is used to make steam that is used to power parts of the factory. This helps cover the cost of running the factory.

▲ This massive chemical factory in Germany produces oleum and sulfuric acid.

NITRIC ACID

Jabir ibn Hayyan is credited with discovering nitric acid, in A.D. 800.

USES FOR NITRIC ACID

▶ Nitric acid is used in the manufacture of the explosives nitroglycerin and trinitrotoluene (TNT), which were both developed in the mid-1800s. These explosives are used to blast rocks apart in mining and quarrying.

▶ Nitric acid is also used in the manufacture of the fertilizers ammonium nitrate and sodium nitrate. These fertilizers are produced by neutralization reactions. This equation shows the reaction that produces ammonium nitrate:

$$\text{Ammonia} + \text{Nitric acid} \longrightarrow \text{Ammonium nitrate}$$

$$NH_{3(aq)} + HNO_{3(aq)} \longrightarrow NH_4NO_{3(aq)}$$

▶ Drug manufacture and the manufacture of synthetic fibers, such as nylon, require nitric acid. Nylon was first developed as a substitute for silk, which is produced by silkworms. Today, nylon is used to make toothbrush bristles, clothing, rope, and carpets.

▶ A more unusual use for nitric acid is to combine it with hydrochloric acid to make aqua regia. This solution is one of the few that is capable of dissolving gold and platinum.

▼ This close-up photograph of Velcro shows a woven base with protruding hooks. A loop catches on the hook. This connects the two surfaces of the Velcro. Nitric acid is used to manufacture nylon, which is used to make Velcro.

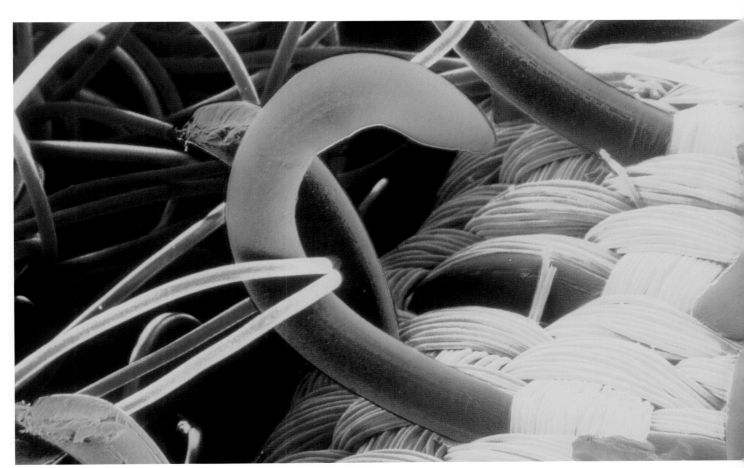

MANUFACTURE OF NITRIC ACID: THE OSTWALD PROCESS

Nitric acid is produced by the Ostwald process. The most important ingredient in this process is ammonia—a product of the Haber process. The Haber process converts hydrogen and nitrogen into ammonia in the following reaction.

Hydrogen + Nitrogen ⟶ Ammonia

$$3H_{2(g)} + N_{2(g)} \longrightarrow 2NH_{3(g)}$$

Most of the ammonia from this process is converted into fertilizers, but a small portion is saved for manufacturing nitric acid. The following steps describe the Ostwald process (nitric acid production).

(1) The production of nitrogen monoxide begins when ammonia is mixed with air and passed through a catalyst made from a mixture of platinum and rhodium. The mixture is heated to 1,652°F (900°C). This reaction produces nitrogen monoxide and water.

Ammonia + Oxygen ⟶ Nitrogen monoxide + Water

$$4NH_{3(g)} + 5O_{2(g)} \longrightarrow 4NO_{(g)} + 6H_2O_{(g)}$$

(2) The mixture of nitrogen monoxide and water is cooled and allowed to react with more oxygen, producing nitrogen dioxide.

Nitrogen monoxide + Oxygen ⟶ Nitrogen dioxide

$$2NO_{(g)} + O_{2(g)} \longrightarrow 2NO_{2(g)}$$

(3) Even more air is added and the nitrogen dioxide is mixed with water. This produces nitric acid.

Nitrogen dioxide + Oxygen + Water ⟶ Nitric acid

$$4NO_{2(g)} + O_{2(g)} + 2H_2O_{(l)} \longrightarrow 4HNO_{3(aq)}$$

Companies that manufacture nitric acid must ensure that no nitrogen dioxide or nitrogen monoxide escapes from the factory, because the gases will dissolve in rain clouds and become one of the ingredients in acid rain. Also, it is essential that no nitric acid goes down the factories' drains. The acid would find its way into rivers and lakes and destroy the wildlife that lives there.

◀ These workers are installing platinum-rhodium sheets, which act as catalysts in the first stage of nitric acid production.

▶ In 1845, a German chemist, Christian Schönbein, was experimenting in his kitchen when he spilled a mixture of nitric acid and sulfuric acid. To clean the spill, he used a cotton apron, which he then hung on the stove to dry. Once the apron had dried, it exploded in a flash. Schönbein had invented an explosive called nitrocellulose, or guncotton. The nitric acid and the cellulose had caused the following reaction:

Nitric acid + Cellulose $\longrightarrow$ Nitrocellulose + Water

$2HNO_3 + C_6H_{10}O_5 \longrightarrow C_6H_8(NO_2)_2O_5 + 2H_2O$

The nitrocellulose contains fuel and oxygen within one chemical compound, which is what makes it explode.

▶ In 1867, the Swedish scientist Alfred Nobel invented dynamite. Nitroglycerin is the basis of dynamite and is made by mixing glycerol, sulfuric acid, and nitric acid. Nobel continued to develop nitroglycerin to make it more stable and, therefore, a useful explosive. To do this, he mixed nitroglycerine with silica and formed it into rods, creating dynamite.

Although his discovery made him very rich, Nobel spent many years feeling guilty about his destructive invention. Over time, laboratory explosions killed his brother Emil and several other people.

Many years later, when Alfred Nobel died, his will contained instructions for his fortune to be used for a series of prizes—the Nobel Prizes. These prestigious awards are given for outstanding discoveries in physics, chemistry, medicine, literature, and peace.

▼ **Dynamite is highly explosive and produces a lot of smoke.**

7

Making gases

Salts are produced from the manufacture of hydrogen gas and carbon dioxide gas. When a metal reacts with an acid, a salt and hydrogen gas are produced. The equation for this reaction is as follows:

Metal + Acid $\longrightarrow$ Salt + Hydrogen

HYDROGEN GAS

In a school laboratory, you can make hydrogen gas by reacting zinc and hydrochloric acid. It is best to use zinc granules because their surface area is large and this increases the rate of reaction.

HOW TO MAKE HYDROGEN

This method of hydrogen collection is called "collection over water." It is used for gases that are lighter than air and not very soluble in water.
(1) Zinc is placed in an Erlenmeyer flask.
(2) Hydrochloric acid is added through a thistle tube. The acid must fill the thistle tube.
(3) A delivery tube is connected to the flask and a trough is filled with water. The other end of the delivery tube is placed under the surface of the water.
(4) Cover the delivery tube with a test tube of water.

The reaction that occurs is expressed in the equation below.

Zinc + Hydrochloric acid $\longrightarrow$ Zinc chloride + Hydrogen

$$Zn_{(s)} + 2HCl_{(aq)} \longrightarrow ZnCl_{2(aq)} + H_{2(g)}$$

The salt remains in the solution. Hydrogen gas travels through the delivery tube into the test tube. As gas fills the test tube, it forces the water out until bubbles appear on the surface of the water in the trough. This indicates the test tube is full. It is important to insert a stopper into the mouth of the test tube while it is still under water.

THE TEST FOR HYDROGEN GAS

To test for hydrogen gas, a lit match is inserted into the tube. If hydrogen is present, it will make a loud, squeaky pop. The squeaky pop is actually a small explosion. Hydrogen is explosive when it is mixed with oxygen.

MAKING HYDROGEN GAS

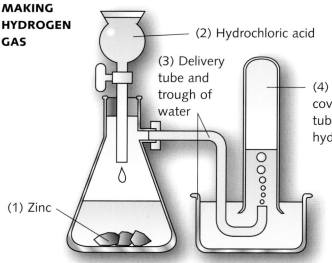

(2) Hydrochloric acid

(3) Delivery tube and trough of water

(4) The test tube covers the delivery tube and collects hydrogen.

(1) Zinc

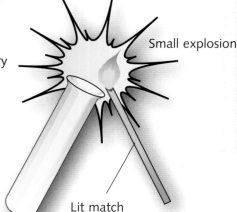

Small explosion

Lit match

HYDROGEN—USEFUL, BUT EXPLOSIVE

Hydrogen is used to make many things from rocket fuel to margarine. Hydrogen is a very light gas. It rises in air, and in the past, it was used in airships. It is also flammable and was probably the cause of the fatal Hindenburg disaster in 1937. Approximately one-third of the people onboard the Hindenburg died when the airship quickly

▲ The Hindenburg airship contained hydrogen fuel. It dramatically exploded as it approached land near New Jersey in 1937.

caught fire. There are several theories regarding the cause of the explosion. Some think that an accumulation of static electricity ignited the hydrogen; others believe that the airship was sabotaged. One theory suggests that hydrogen leaked from a hole caused by a snapped wire and was ignited by a spark of static electricity. Another theory asserts that it was the covering of the airship—rather than the hydrogen—that provided the fuel.

CARBON DIOXIDE GAS

Carbon dioxide is used in fire extinguishers, carbonated beverages, and as stage smoke. If carbon dioxide is cooled, it turns directly into a solid without passing through the liquid stage; this is called **sublimation**. Stage smoke is produced by the sublimation of solid carbon dioxide. Stage smoke is also called dry ice.

HOW TO MAKE CARBON DIOXIDE

Carbon dioxide is produced by the reaction between an acid and a metal carbonate.

Calcium carbonate + Hydrochloric acid $\longrightarrow$ Calcium chloride + Carbon dioxide + Water

$$CaCO_{3(s)} + 2HCl_{(aq)} \longrightarrow CaCl_{2(aq)} + CO_{2(g)} + H_2O_{(l)}$$

Although there are three products and two reactants, only one of these is a gas—carbon dioxide. The same equipment that is used for hydrogen collection is also used for carbon dioxide collection.

THE TEST FOR CARBON DIOXIDE

When carbon dioxide bubbles through calcium hydroxide (limewater), the mixture becomes milky. This is because insoluble calcium carbonate forms.

Glossary

ACID – A substance that dissolves in water and releases H^+ ions (protons). Acids have a pH less than 7.

ACID RAIN – Rain that contains high levels of nitric acid, sulfuric acid, or both. Acid rain has a pH of approximately 5. Acid rain forms when gases from burning fuels combine with moisture in the atmosphere.

ALKALI – A solution formed by dissolving a base in water. This causes the release of OH^- ions. Alkaline solutions have a pH greater than 7.

BASE – The oxide or hydroxide of a reactive metal, such as sodium. Bases can neutralize an acid and produce a salt. Bases can be soluble or insoluble.

BUFFER SOLUTION – A solution that resists changes in pH when small amounts of acid or alkali are added.

BURETTE – A cylindrical piece of glassware used to measure the volume of a liquid and dispense the liquid from a tap on its bottom end.

CATALYST – A substance that accelerates a chemical reaction without itself being chemically changed.

CORROSIVE – A chemical that destroys metal, body tissue, or both.

DISSOLVE – When a solute goes into solution with a solvent.

DOUBLE DECOMPOSITION – A reaction during which two compounds swap chemical partners.

END POINT – The point at which an exact amount of alkali has been added to an acid (or vice versa) for complete neutralization to have taken place.

EVAPORATE – A physical change from a liquid to a gas.

FILTER – The separation of a solid from a liquid by passing the mixture through filter paper.

HYDRATED SALT – A salt that is bound to water.

ANSWERS

Page 7: Investigate
(1) The bone will be smaller and pitted because the acid in the vinegar has dissolved the calcium.
(2) The egg should go all the way in. The vinegar has dissolved some of the shell, making it softer.

Page 11: Investigate
Green/yellow = alkaline
Red = acidic

Page 13: Test yourself
(1) Vinegar = Orange/yellow, pH 3
(2) Ammonia solution = Violet, pH 11
(3) Pure water = Green, pH 7

Page 22: Test yourself
You would see the formation of an insoluble white precipitate of silver chloride. The two aqueous solutions would form silver chloride and a solution called ammonium nitrate.
$$NH_4Cl_{(aq)} + AgNO_{3(aq)} \rightarrow AgCl_{(s)} + NH_4NO_{3(aq)}$$

Page 27: Test yourself
(1) Calcium and sulfuric acid
$$Ca_{(s)} + H_2SO_{4(aq)} \rightarrow CaSO_{4(s)} + H_{2(g)}$$
(2) Copper and diluted nitric acid
$$3Cu_{(s)} + 8HNO_{3(aq)} \rightarrow 3Cu(NO_3)_{2(aq)} + 2NO_{(g)} + 4H_2O_{(l)}$$
(3) Magnesium and hydrochloric acid
$$Mg_{(s)} + 2HCl_{(aq)} \rightarrow MgCl_{2(aq)} + H_{2(g)}$$

Page 27: Investigate
The iron nail will look corroded. The iron has reacted with the acid and formed a salt. You may also see bubbles of hydrogen gas.

Page 29: Test yourself
You would mix together magnesium oxide and sulfuric acid. You would then filter the mixture, heat the solution to evaporate some of the water, and create a saturated solution. You would then leave the sample to crystallize. You would not use the above method to obtain potassium sulfate crystals because potassium oxide—one of the reactants you would need to use—is soluble.

Page 31: Test yourself
To make potassium chloride, put a known volume of one of the reactants into an Erlenmeyer flask. Add an indicator to the beaker. Put the other reactant in a burette. Gradually add the reactant from the burette to the flask. When the indicator changes color, neutralization has occurred and potassium chloride has formed. Note the amount of reactant added. Then repeat the process without the indicator so that your final product is not colored by the indicator. Add the same amount of

INDICATOR – A substance that changes color depending on the pH value.

INSOLUBLE – A substance that will not dissolve in a given solvent.

ION – An atom that has a charge.

IRRITANT – A substance that causes inflammation on the skin. Irritants usually cause itching when applied to skin.

LITMUS – A type of indicator.

NEUTRALIZATION – The reaction of an acid with an alkali that produces a salt and water.

ORGANIC COMPOUND – A compound that contains carbon (except carbon dioxide, carbon monoxide, and carbonates). Organic compounds can also contain hydrogen, nitrogen, and oxygen.

OXIDIZING AGENT – A chemical that easily releases its oxygen.

PIGMENT – An insoluble colored substance.

PIPETTE – A piece of glassware used to measure and transfer one volume of liquid.

PRECIPITATE – When a solid separates from solution by chemical or physical change.

SALT – A compound made of chemically bonded positive and negative ions.

SATURATED SOLUTION – A solution that cannot dissolve any more solute.

STANDARD SOLUTION – A solution of a known and exact concentration.

SUBLIMATION – A physical change from solid to gas, or vice versa.

SUSPENSION – A mixture in which fine insoluble particles are dispersed in a liquid.

TITRATION – A chemical method used to determine the concentration of an acidic or alkaline solution.

UNIVERSAL INDICATOR – A mixture of indicators. A universal indicator changes color across the whole range of pH values.

VOLUMETRIC FLASK – Laboratory glassware that accurately measures one volume of solution.

WEIGHING BOAT – A device used to measure the mass of a substance. It is a boat-shaped dish that sits on weighing scales. The dish is removable so that it can be easily washed.

---

reactant that you did in the first part of the experiment. Evaporate the solution to a saturated solution over a water bath. Then leave the apparatus near a sunny window for one week. Cover the dish with filter paper to protect the solution from dust.

Page 33: Test yourself
Marble is a metamorphic rock that is made of calcium carbonate. Marble is not porous. This means that it does not contain tiny holes. Therefore, it does not allow rainwater to enter and weather the rock. Limestone is a porous form of calcium carbonate. It is easily weathered and eroded by rainwater.

Page 33: Investigate
There is a fizzing sensation in your mouth caused by the formation of carbon dioxide. The reaction is expressed by the following equation. Citric acid + Sodium hydrogencarbonate ➞ Sodium citrate + Carbon dioxide + Water

Useful Web sites:
http://www.chem4kids.com
http://www.howstuffworks.com
http://www.chemtutor.com/acid.htm
http://library.thinkquest.org/3659/acidbase/

Index

Page references in *italics* represent pictures.

PHOTO CREDITS – *(abbv: r, right, l, left, t, top, m, middle, b, bottom)* **Cover background image** www.istockphoto.com/Mike Bentley **Front cover images** (r) www.istockphoto.com/Joshua Haviv (l) www.istockphoto.com/vera bogaerts **Back cover image** (inset) www.istockphoto.com/vera bogaerts **p.1** (t) www.istockphoto.com/Daniel Slocum (br) Charles D. Winters/Science Photo Library (bl) www.istockphoto.com/oana vinatoru **p.2** Dale C. Spartas/Corbis **p.3** (tr) www.istockphoto.com/Pattie Steib (b) www.istockphoto.com/Iurii Konoval **p.4** (tr) www.istockphoto.com/Ufuk ZIVANA (tl) www.istockphoto.com/Anka Kaczmarzyk (br) Charles D. Winters/Science Photo Library **p.5** www.istockphoto.com/Joshua Haviv **p.6** www.istockphoto.com/Jeremy Voisey **p.7** (b) Biophoto Associates/Science Photo Library (t) www.istockphoto.com/Andriy Doriy **p.8** www.istockphoto.com/Anka Kaczmarzyk **p.9** Cordelia Molloy/Science Photo Library **p.10** www.istockphoto.com/Paul Cowan **p.11** (bl) Andrew Lambert Photography/Science Photo Library (tr) www.istockphoto.com/Daniel Slocum (br) www.istockphoto.com/Vera Bogaerts **p.12** Andrew Lambert Photography/Science Photo Library **p.15** Dr Jeremy Burgess/Science Photo Library **p.17** (t) Andrew Lambert Photography/Science Photo Library (b) Andrew Lambert Photography/Science Photo Library **p.20** www.istockphoto.com/Joshua Haviv **p.21** www.istockphoto.com/Weldon Schloneger **p.22** Martyn F. Chillmaid/Science Photo Library **p.23** Sinclair Stammers/Science Photo Library **p.24** www.istockphoto.com/Alan McCredie **p.25** (t) Dale C. Spartas/Corbis (b) Charles D. Winters/Science Photo Library **p.26** Charles D. Winters/Science Photo Library **p.27** Charles D. Winters/Science Photo Library **p.30** Richard Megna/Fundamental Photos/Science Photo Library **p.31** Peter Menzel/Science Photo Library **p.32** Martyn F. Chillmaid/Science Photo Library **p.33** www.istockphoto.com/Pattie Steib **p.35** (l) Charles D. Winters/Science Photo Library (r) Burstein Collection/Corbis **p.36** www.istockphoto.com/Anka Kaczmarzyk **p.37** Science Photo Library **p.38** David Taylor/Science Photo Library **p.39** (b) www.istockphoto.com/oana vinatoru (t) www.istockphoto.com/Ufuk ZIVANA **p.40** Mathias Ernert/epa/Corbis **p.41** Dr Jeremy Burgess/Science Photo Library **p.42** James L. Amos/Corbis **p.43** www.istockphoto.com/Iurii Konoval **p.45** Bettmann/Corbis